First cultivated in ancient China this protein-packed, vitamin rich, king amongst beans is unmatched nutritionally by any other food. Its story is told and its life-giving secrets proclaimed in this book.

ABOUT SOYA BEANS

Wonder Source of Protein and Energy

by

G.J. BINDING
M.B.E., F.R.H.S.

THORSONS PUBLISHERS LIMITED
Wellingborough, Northamptonshire

First published 1970
Second Impression 1971
Third Impression 1977

ISBN 0 7225 0147 1

Printed and bound in Great Britain by
Weatherby Woolnough, Wellingborough,
Northamptonshire

CONTENTS

ACKNOWLEDGEMENTS

I wish to express my sincere thanks to Messrs. S. Puspalm, K. Menon, John Pang, Haji Sulieman, and other friends (and their families) in the Far East, through whose hospitality my family and I first encountered the Soya bean in its many forms.

I also wish to thank:

The President, Soybean Council of America, Washington D.C., U.S.A.

The Editor, The Soybean Digest, Hudson, Iowa, U.S.A.

Worthington Soy Foods Inc., Worthington, Ohio, U.S.A.

CHAPTER ONE

BEANS IN GENERAL

THE BEAN, ONE of the oldest vegetables known to man, has always been surrounded by strange stories, legends and sayings. We are all familiar with that lovable character Jack and The Beanstalk, and such expressions as 'Full of beans'. Having a good time was often referred to as a 'Beanfeast' or 'Beano'. The person who presided as King over the 12th Night Festivities was named the 'Bean King'. He gained this honour by getting the bean buried in the 12th Night Cake.

Types of Beans

There is hardly another vegetable which produces so many different varieties, colours and shapes as the bean. World wide there are no less than 150 species of endless strains and every hue of the rainbow. Beans have always been recognised as delicious, appetising and wholesome, full of body building goodness. One cannot obtain a finer vegetable than freshly gathered beans cooked and served straight from the garden. I have never met anyone young or old, who does not like at least one kind of bean. This vital food is taken so much for granted. During winter months when crops cannot be grown, frozen beans are available and a wide choice of tinned kinds to satisfy all tastes. For those who dislike frozen or canned foods there are numerous dried beans until fresh vegetables are again in season. So at all times beans can be obtained to provide meals; from bacon and beans at breakfast, an odd snack of beans on toast, or as a fine health giving supplement to other vegetables at a main meal.

Folklore

Beans play a conspicuous part in the Folklore of England and many other countries. In Yorkshire, even today some believe that sleeping in the beanfields will produce vivid dreams. Miners, in many places still believe that accidents in pits are more likely to occur when beans are in flower. Heathen peoples always considered that plants with strong smelling flowers, especially the bean, were closely related to ghosts and the deceased. A Roman festival took placed on the 1st June in which beans were offered to the dead. An ancient Aryan traditional religious belief was that beans and honey were held to be sacred food to be offered to the departed. This little seed was also used in a form of balloting by the Romans and Greeks when a white bean signified Yes and a black one No. In Italy until about 50 years ago it was the custom to distribute beans to the poor on All Souls Day. A Josh Billings is impressed by the fact that one quart of beans on being boiled for 2 hours came out as one and a half gallons. A very strange account was written by Pliny, the Naturalist, that man would emerge from manure heaps if beans were planted within them.

The Soya Bean

So much may be said about beans available to man today, what he thinks about them now, and the strange beliefs and customs surrounding this vegetable in an age gone by. However, one little specimen of the bean family, conspicuous by its absence as part of the diet of the average English Family, is the soya bean. This is to be deplored, and to say the least is unfortunate.

Prior to the second world war England was one of the world's largest importers of the soya bean. This state of affairs no longer exists today. The decline in popularity of this vegetable, in England and on the continent, may be influenced by the fact that the crop cannot be successfully cultivated in Europe. Furthermore, people today in times of full employment and greater prosperity seek more expensive processed foods, rather than natural health giving vegetables. As a result a fine nutritious food which gives 100% value for money, and is cheap enough to be within the reach of all is lost.

More Value for Money

Josh Billings, the character out of Folklore, claimed that one quart of beans became one and a half gallons on boiling for 2 hours. This, to say the least, is a great exaggeration. However, from my own vast experience of the soya bean, one pint, will produce, for a family of four, a daily vegetable supplement, served with other vegetables at one meal per day for about a week. Like the procedure reported by Josh Billings, our pint of beans (dried) are first soaked in water overnight, brought to the boil, and allowed to simmer for about 30 to 45 minutes. The least amount of boiling the harder the bean and the more nutty the flavour. There are many ways of serving and preparing them, given in detail in the chapter on recipes. So, in spite of the incredibility of the statement of Billings, our pint of dried soya beans purchased for about 1/8 today is without doubt one of the most economic foods available. It may be said that it is the cheapest food on the market, if you can obtain it, and yet the most health giving. It is hoped that this little booklet will introduce strangers to the bean, stimulate their interest in this fine vegetable, and tempt them to try it. Others, who are conversant with the bean, and eat it in one or more of its numerous forms, as part of their diet, may learn more of its versatility, wonderful health giving properties, and realize what a wise choice they have made.

Wonder Food - God's Gift to Man

From time immemorial man has searched the wide world for everlasting youth and a perfect food. He, even in this atomic age, is almost as far away from success as ever in both these probes, except for the fact that he has the finest, if not the most perfect food, on his doorstep, if the Far East and places like the United States can be considered as being so close in the Jet Age. Credit for this food goes to the ancient people of the Orient, who were the first to cultivate and proclaim the vital life giving secrets of the soya bean. Without exception it is, as yet, the most perfect food known to man, ever grown, and most unlikely to be excelled in the future. Our soya bean, although very small in size, is in protein, vitamin and mineral content undoubtedly the King of all Beans. When it was first introduced into America it became known as the Japan Pea. For dried in its natural shape it is not a bit like a bean but is perfectly round, even smaller than a pea. Of a light yellow colour it has a tiny black dot upon it. After being soaked for 12 hours in water, the size is more than doubled and it takes on the shape of an ordinary dwarf bean. For over 5,000 years this tiny seed has been the staple food of certain parts of the East, including North China, Japan, Korea and some areas of India. The ancient Yogis, who were among the world's first vegetarians, placed great faith in the soya bean as a supplement to their meatless diet. It fact it became known as, and still is in the Far East referred to as being 'The Meat of the Soil'. If you read on, it is hoped that this short account of the bean will show you that it not only earned such a high place in the minds of the ancient peoples of the Far East and India but even today in our 20th Century the tiny little bean is still well and truly The Meat of the Earth.

Many Valuable Foods and Essentials in one plant

The soya bean is a wonderful source of vital goodness. Nutritionally, there is no other food that compares with it. It is the only vegetable which gives complete protein which compares with, and is in fact identical to, all forms of animal meats. It is really as good as eggs, milk and any other protein food. In fact superior in many ways. (The writer is not a vegetarian, but only a moderate meat eater). The protein content alone of the bean is enormously high, being over 35%. It has 18% fat and is very rich in vitamins, trace elements, copper, iron and other ingredients vital for the development of healthy bones and bodies. This is not the end of our story of this beanfeast for its crowning glory lies in the fact that it is rich in lecithin, plant sterols and unsaturated fatty acids. Lecithin plays a most important part in lowering the amount of cholesterol in the blood stream, thereby helping to prevent coronary heart disease and weaknesses of the arteries. Fuller details of the bean's vitamin, mineral and other qualities are given in a later chapter.

CHAPTER TWO

HISTORY OF THE SOYA BEAN

As STATED, THE little bean was shown to have been first cultivated in the Far East long before recorded History. Records have been traced in China giving details of planting of crops over 2,000 years before Christ. Advice to growers in those ancient times on soil conditions, types of crops and cultivation hints were found all in great detail. Just like the present day farmers guide to the planting and rotation of grain crops and so on. All this was taking place in times when the modern methods of agriculture were beyond anyone's wildest dreams.

Cultivation and Importance to the East

Grown as an annual crop it is without doubt the most important legume cultivated in Japan, Korea, North China and parts of India. Many Indian castes are vegetarians and the cow is considered a sacred animal and not allowed to be slaughtered under any circumstances. Perhaps this is why the Almighty allowed a bean equally as nourishing to flourish in this part of the world. For this and other reasons, to many of the varied and differing races of the East the soya bean is not simply taken for granted as a vegetable but has earned that title 'the meat of the soil'. Those ancient races of China and Japan were so impressed and thankful for their crops of soya beans that it was held in very high esteem, and in fact was classed as being one of the 'Five Sacred Grains'. Such crops were not only considered sacred and being in every way vital to the health, but essential to the existence of these ancient civilisations. So it will be appreciated what regard these people had for the soya bean then, and still have today, 5,000 years later. It has withstood the test of time from one generation to another and plays just as much a vital role now as it did then. The four other cultivations which made up the the 5 sacred grains were those of barley, rice, wheat and millet.

Many useful inventions, customs and traditions have been passed down from the wise Chinese and other Eastern Races. Without doubt, although little realized, the greatest contribution they have made to the world in regard to food, is the production and continuous cultivation of the fabulous little soya bean. They were always aware of the fact that the bean is almost 'a miracle food'. Unfortunately, people in many other countries, even if they know of the existence of this crop are not prepared to accept something which is predominantly Eastern in origin. Little do they realize what they are missing in food value, but I hope this booklet may enlighten some of them.

Other Areas of Cultivation

Attempts to cultivate the bean in Europe have not met with great success. It is a loss to the people of these countries that this crop will only really do well in warmer climates. However, in more recent years it has been grown with some success in Russia, and the Americans have made tremendous advances, being today one of the largest producers of soya beans in the world. Between the two wars a flourishing industry revolving round the soya bean has sprung up in the U.S.A. Attempts have also been made with varying results to cultivate the bean in Australia, New Zealand, Manchuria, The Philippines, Formosa and Thailand.

Soya Beans in Europe

Credit for bringing the news of the soya bean to Europe goes to a German botanist Englebert Kaempfer. Towards the latter part of the 17th Century he spent some three years in Japan and described the bean as being similar to a lupin with seeds of a pea. Interest in the crop did not materialize in Europe for nearly another 100 years. In 1875, successful crops of different varieties of the bean were cultivated in Austria by Fredrich Haberlandt. The result was that four good strains appeared in Austria, Hungary and other countries and reasonable crops were grown.

Haberlandt's efforts did much to promote interest in the growing of the bean. In 1739, packets of the seed were taken to France by Missionaries from China, but it was not cultivated for another 40 years, when some Chinese soya beans were grown in the Marseilles area. Likewise, Italians were able to plant the bean in about 1840 but it was never cultivated as a major crop. This is unfortunate, the Italians, undoubtedly have the ideal warm climate in which it would flourish. An attempt, with limited success was made to grow it at Kew Gardens, England. The English climate is far too cold for the crop to do well.

Ignorance of the True Value of the Bean

During these early days of crop cultivation in Europe, and unfortunately even today, people considered the bean to be unfit for human consumption. No doubt many thought it to be an ideal food for Chinese coolies, or Rickshaw boys. What a feast it was for them—how they could run with a rickshaw all day long, or work like beavers refuelling a ship with heavy baskets of coal with no ill effects. Others, farmers included, felt that the bean was ideal as cattle fodder and equally as good if sown, allowed to grow and then ploughed back into the soil as a green manure. This is, of course, really rather tragic, but it must be remembered, only in America and Europe in comparatively recent years, scientists and doctors alike have come to realize the necessity of vitamins to health, together with a reasonable amount of minerals and protein. Furthermore, it is only during the last few years of investigation into stress, heart diseases and other illnesses of modern society, that the real value of the soya bean has been revealed. Once again, I will mention from nearly 20 years' personal experience of the soya bean as a food—I have found it to be second to none.

History of the Soya Bean—U.S.A.

In his writings in 1840, a certain James Mease reported the bean to be flourishing in the Pennsylvania area of America. Later, in 1854, some varieties were brought to the U.S.A. from Japan—and at that time it very appropriately became known as the Japan pea. For many years which followed the bean was simply used by farmers in America as food for livestock, or ploughed into the fields as manure. The Perry expedition to Japan brought back strains of the bean to the U.S.A. in approximately 1854, with the ideal climatic conditions prevailing the crops were very good. It was in 1882 that the Carolina Experimental Agriculture Station grew what became known as the Mammoth Yellow Soya Bean, which proved an excellent cropper and ideal for growing. In 1889 a representative of the Massachusetts Agriculture Experimental Station introduced a large number of species of the bean into the States. The Kansas Agricultural Experimental Station likewise carried out tests and experiments on large scale growing of the crop. Realizing the possibilities for American farmers, the U.S. Department of Agriculture introduced large numbers, in fact over 10,000 species of soya beans, mostly from Asia, with a few varieties from Europe.

By 1936 the Americans were well on the way to becoming the world's largest growers of the soya bean. The U.S.A. and the Canadian Departments of Agriculture co-operated to produce better and improved varieties of seed. Vast amounts of work and experiments conducted resulted in bigger and better crops of beans, with correspondingly greater yields of soya bean oil. The value of the crops increased annually and a new industry was formed.

Major American Industry

During the past 50 years the Americans have become one of the world's largest growers of the soya bean. It is indeed remarkable, in fact almost unbelievable, to what uses, both as a food and a drink, and for industrial purposes the soya bean has been put to in the U.S.A. This does not include uses on the farm as manure and fodder for animals. A further booklet could well be written on the commercial aspects of the bean in the States. However, as we are looking at it from the food value, a brief account of its other very numerous uses will be given in a later chapter. To me it seems a criminal waste of a perfect food to utilize soya beans to further man's end in the manufacturing world, especially at a time when millions are on the verge of, if not exactly, starving.

CHAPTER THREE

CONTENT OF SOYA BEANS

WITH AN ABUNDANCE of vital bodily needs, soya beans contain over 40% protein and about 20% fat, or bean oil. This vegetable oil is ideal as it does not cause any great increase in consumers' body weight, like animal fats. Proteins are made up of amino acids, of which there are 22 in all. The soya bean is one of the few foods that contain all 22 of the health giving amino acids. As such it is considered a complete protein food. These beans are very rich in lecithin, a wonderful protector of the heart and arteries (amino acids and lecithin are discussed in greater detail in Chapter IV). Soya bean sprouts, or bean shoots, as they are sometimes called, contain vitamin A and are rich in vitamin C. Likewise, the bean is a good source of vitamins B and E. Minerals and trace elements are also present, these include copper, iron, calcium, magnesium, zinc, manganese, nitrogen, phosphorous, potassium and sulphur.

The rich supply of protein, vitamins and minerals build strong healthy bones and help protect the body from disease and sickness. All these ingredients of the bean play a fundamental part, as we shall see, in the efficient functioning of our bodies.

Protein

What is it? The word means primary, or first place and this is what it is to human beings and animals. We are made up of protein and every part of our body contains it. Body cells make up tissues, and each and every one of the millions of cells relies on a supply of protein to counteract daily wear and tear, and maintain good health. The body is able and can store a limited amount of protein, but this is quickly used up if daily supplies cease, or are reduced to a low level. The main function of protein is body repair and to generate heat and energy. Naturally,

17

persons who are overweight need to control their daily intake of protein. Others who are thin can beneficially add protein to their diet. As is the case with all foods and vitamins, to consume excessive amounts may well do more harm than good. Moderation in all things is the way to healthy existence, and eating is no exception to this rule.

Amino Acids

It was not realized for a long time that various proteins are made up of nitrogen, and up to 22 amino acids. Most body cells require all these for perfect development and functioning. When protein is consumed it is broken down by the body into the various amino acids which are carried by the blood stream to every cell, this process being continued as long as we live.

Wonderful Functioning of the Body

A remarkable thing is that the cells of the body are able to produce 14 of the 22 amino acids. Protein, fat, sugar and nitrogen are used for this purpose. However, there are 8 amino acids which the body cannot manufacture, which are sometimes referred to as being 'the essential amino acids'. In order to enjoy good health it is vital that these are present in our daily diet. From this it will be appreciated how important it is to have not only sufficient protein, but a reasonable amount which contains all the 22 amino acids.

Foods Rich in Protein

Many foods contain a good supply of protein including: meats, liver, milk and all milk products, yeast, millet, sunflower seeds, cereals, nuts and other seeds, including the richest source of vegetable protein, the soya bean and its products. Foods with all the 22 amino acids, that is, complete protein, are meats, eggs, liver, milk, brewer's yeast, wheat germ, nuts and soya beans.

It is interesting to note that the amino acid analysis of soya beans and casein (composition of meat) are almost identical in every way. The only food other than meat which has anything like the protein content of soya products is skimmed, dried milk. Half a cup of skimmed milk is as nutritious as a quart of fresh milk. The reason for this is quite simple for all the essential ingredients of the milk powder are present in the skimmed part and not contained in either the cream or water.

Lack of protein

In spite of being able to store some protein, if the body does does not receive an adequate daily supply, eventually ageing will surely occur. If this state of affairs continues with modern stress, muscles will eventually begin to lose tone and degeneration of the body will set in. All manner of illnesses can be attributed to lack of sufficient protein over the years. Poor circulation, chilblains, constipation. A rather dangerous situation can develop because protein deficiency over a long period can be concealed by eating large amounts of devitalized foodstuffs, such as sugars, processed refined foods, white flours and starchy preparations. But the time will come when the body can no longer continue to function properly without the recuperative intake of protein, and lack of it will result in illness and disease. The supply of protein in the body can easily be depleted by persons who worry and suffer anxiety. Those to whom the stresses and strains of daily living are too great, need far more protein than the characters who are carefree, unruffled, and can take life in their stride. Even today a surprising number of diets are sadly lacking in sufficient protein. So in order to remain healthy, young and full of vitality it is essential to obtain enough protein daily. Every part of you is made up of protein, and there is no easier, cheaper or tastier way of getting it than by consuming the soya bean and its products.

Lecithin

If the vigour and vitality that made you feel that it was good simply to be alive has gone, it may well be your body lacks lecithin. This is an ingredient of soya beans and wheat germ which plays a vital part in building up nervous energy in your body. It contains other substances which help prevent fatty deposits being formed in the arteries and other places. The bean has so many things that are good for us but lecithin is that little something extra which places it on its own as a king above all all beans. A more detailed account of the way in which lecithin helps prevent ageing and restores lost youth follows in the next chapter.

Vitamin A.

An adequate supply of this vitamin counteracts infectious diseases. Shortage will cause all manner of complaints affecting the breathing system, bronchial tubes, and even the lungs. It can very soon cause defective eyesight if persons are deprived of it for a long period.

Vitamin C. (*Asorbic Acid*)

Bean sprouts are very rich in this essential vitamin. Acute shortage of it will lead to scurvy, as experienced by sailors before faster steamships were built. It is essential to obtain a regular supply of vitamin C because the body is unable to store it. Signs that a person is not getting a sufficient intake are tiredness, despair, poor healing flesh, and bleeding gums.

Vitamin B.

Today far more is being learnt about the food content of the B vitamin group. It has now been established that, including vitamin B12, there are no less than 11 vitamins in this group. Five of the eleven are present in soya beans, namely, thiamine, riboflavin, niacin, choline and inositol. Of all these 11 vitamins, three are considered most important for the body—the bean contains these three, being the first three mentioned—thiamine, riboflavin and niacin.

We will very briefly examine these three and see what they will do for us.

Thiamine

Sometimes called aneurine. Shortage results in stunted growth and depression. If it is withheld for a long period, beri-beri might occur. It feeds the nerves, aids appetite, and can prevent a large number of minor disorders, such as constipation and other complaints, which, if neglected for years, will cause serious illness.

Riboflavin (B2.)

This ingredient of the bean protects the muscles of the eyes, the skin, hair and finger nails. Shortage will result in cracked lips and corners of the mouth. The tongue may become sore and a burning sensation is caused. Women are more prone to a shortage of riboflavin.

Niacin (Nicotinic acid)

This is sometimes known as the antipellagra vitamin. A serious deficiency of niacin will produce mental problems. Signs of a shortage are: rough skin, sore tongue and diarrhoea. The entire mucous membranes of the mouth, nose and throat are effected.

Vitamin E

This is sometimes known as the fertility vitamin. In recent times vitamin E has proved effective in healing various heart complaints. Such illnesses as thrombosis, phlebitis and ulcers have responded well to vitamin E treatment. It has also been used successfully to cope with menopause problems, to improve muscle tone and overcome poor circulation.

Copper

A component of the bean, and another mineral seldom considered by the average person to have any effect on our bodies.

People are aware that iron is vital for a rich blood stream, and calcium essential for strong teeth and bones, but know nothing about copper. It aids the body's digestive processes and carries out vital work, though not in such a spectacular way, of helping convert iron into hemoglobin (the oxygen supplying substance in red blood cells). This process is vital for the prevention of anaemia. Lack of sufficient copper in the body may well cause faulty respiration and general bodily weakness.

Iron

Undoubtedly one of the most important minerals needed by man. It controls the body's built-in thermostat, ensuring that we are warm in cold weather. More important, it gives vitality, strength, endurance, and staying power. In fact it helps us to obtain from life all these things which are worthwhile, lack of which brings misery and unhappiness. So iron is another vital contribution of the bean.

Calcium

Is really a wonderful food element of the soya bean. Vital in man for the development of strong healthy bones and teeth. Experiments among animals given calcium shows that they live longer, and retain sexual powers well into old age. Whether humans receive the same benefits remains to be proved.

Fat Content

About one fifth of the bean is composed of fat or if reduced, to soya oil. It is of interest that this fat, or oil, is immediately assimilated by the body during daily activity, during the course of repair to vital organs, nerves and cells. So fortunately we do not store the rather high fat content of this vegetable, as we would animal fat. Thus, with its massive supply of protein and fat that will not induce obesity the bean is equally good and nutritious for the under and over-weight consumer.

Vegetarians and the Soya Bean

Whether or not you decide to become a vegetarian, is indeed a personal choice. It may depend on religious convictions, conscience and so on. However, much depends on climatic conditions. One can hardly expect an Eskimo to adopt a vegetarian diet, living in a barren waste, with temperatures well below freezing all the year round. The intense cold of the Arctic and other places make increasing demands upon the constitution, which many feel can only be satisfied on a diet rich in meat. In semi-tropical places and on the equator, the demand for meat dishes is considerably less. Vegetarian foodstuffs are much more satisfying in a hot climate, so it is fitting that the soya bean flourishes under these conditions, where it is in greatest demand. Unlike meat eaters, it is a far greater problem for vegetarians to ensure they get an adequate daily intake of protein. A regular supply of vegatables and drinks rich in protein is important. The most well-known protein foods have been mentioned, but for the benefit of vegetarians, the three essential foods are soya beans and the product of the bean, skimmed milk powder and nuts, in particular brazil nuts. Throughout the ages the bean has fully earned its description as 'the meat of the soil' and as such it is God's gift to everyone, but in particular to those following a meatless diet. In fact, by replacing meat with soya beans, and bean products, vegetarians are getting their protein direct from the soil, whereas meat eaters are really making use of second class protein.

Vegans

For followers of the Vegan diet, which excludes both meat and eggs, I would suggest that the soya bean and as many of its products as can be obtained is a *must* and should be eaten daily without fail.

Future for the Hungry People of the World

Milk was always accepted as being the most perfect food for human beings. However, in soya beans we have a variety of nourishment which excels anything ever produced in the past. America is growing over 60% of the world crop of beans, and over one billion bushels were harvested in 1968. An even larger crop was forecast for 1969. Undoubtedly, such a supply of health giving food, which can be utilized in many different forms, should prove a Godsend to prevent starvation among the millions living in underdeveloped countries.

CHAPTER FOUR

ABOUT LECITHIN—VITAL FOR RETAINING YOUTH

IN MANY COUNTRIES today people have every prospect of living longer, and of being able to remain active well into old age. Your date of birth is not now the only, or most deciding factor of when you are old. There are renowned examples of men and women who have defied the calendar. Winston Churchill became Prime Minister at 65. Conrad Adenauer, leader of post-war Germany, known to the people as 'Der Alte', was leading a great nation at the age of 85. Other more recent events include the brave exploits of men like Chichester and Alec Rose, who at 65 years of age sailed round the world single-handed. In Western civilization the population of countries becomes an older one, there are more people than ever before living to retirement age and a remarkable number who now complete their century. Not many years ago it was unusual to meet a person of 100 years of age—not so today. However, in spite of this extensive increase in life span, many unfortunate people do age quickly, for example deaths from heart disease and cancer have never been so high. Certain factors do contribute towards life expectancy—those whose parents and grandparents lived to a ripe old age can reasonably expect to do likewise. However, in spite of setbacks such as these which are beyond our control, we are to a far greater extent than many realize, masters of our own destiny. 'We are today, what we were yesterday, and we will be tomorrow, what we are today.' Time does not stand still. Our bodies can only be made up of what we put into them in the form of food and drink. Thus, to spend years consuming devitalized processed foods can eventually produce a human body, tired and weary, as equally deprived of vitality as the unwholesome diet received. To be healthy, and remain in good physical condition, it is imperative to eat wholesome, natural food as God provided it.

Plenty of fresh fruits and vegetables, whole meal flour products, little or no refined sugars. Such sweeteners as honey, fruit juices and natural brown sugars are by far the best. Many pay far more attention to the running of a car engine, or lawn mower, than to their own bodies. Inferior oils and misuse will soon ruin an engine. Fortunately, devitalized feeding, even over a long period will not permanently turn anyone into a weakling or chronic invalid. The human body, even after gross neglect, has a wonderful way of responding to correct food and healthy living. In recent times, scientists, doctors and others have shown that approximately every 7 years for a long period of our lives practically every part of the human body has been replaced. So it is never to late to mend. An almost miraculous change can be brought about by getting nourishing food and other essentials that make life worth living.

What is Lecithin?

It is a food substance, rich in choline and inositol, which is derived from certain foods, such as egg yolk, wheat germ, and by far the richest source—soya beans. Apart from an abundance of protein, and all the other goodness, lecithin is that little extra ingredient which makes the soya bean superior to any kind of food known to man. When those ancient people christened the bean 'the meat of the soil', they were thinking only in terms of a satisfying food which would more than replace vital protein elements of a meat diet. Little did they dream about lecithin, for it was a long time indeed before the evidence of the presence of this substance in the crops was revealed. Lecithin is utilised by the human body to build up much of the tissues of nerves and brain. About one third of our brains and a fifth of our nerves are composed of it. In our nervous system it is used to produce electricity. Lack of this substance in our diet will eventually lead to nervous exhaustion, for it is actually an ingredient of nerve tissues. For many, who lack enthusiasm for living, lecithin might well prove to be the difference to them of being alive—or just existing. So, in simple terms lecithin is really a substance which contains generating power for the smooth running of the body, just as an effective dynamo will keep an engine running.

Cholesterol

This is a fatty kind of residue that builds up in the bloodstream and causes arteries to become furred up. It is believed to be contained in animal fats, but even experts have differing opinions of cause of cholestrol. It is very interesting to note, however, that strokes and heart diseases increase directly in proportion to a country's industrial development. Western civilization has a high rate of death from strokes. The poulations of these areas are able to, and do consume a plentiful supply of calories, dairy produce, fat, together with a high intake of refined sugar, white flour and all manner of processed foods. Where people are living on more natural foods there is a far less incidence of strokes. Vegetarians show lower incidence of strokes than usual. Generally, the greater the physical activity, the less chance there is of getting heart disease and strokes. This is clearly shown in U.S.A. and England—where professional and clerical workers have the greatest risk and agricultural and manual workers the lowest. To give some idea of the increase in deaths from strokes the figures of the Register General records are quoted.

1944 nearly 50,000 died of strokes.

1960   „   66,000   „   „   „

When it is appreciated that nearly two-thirds of those who suffer a stroke recover, a clearer picture is revealed of how serious and widespread this illness is becoming.

Choline

This component of lecithin aids in the prevention of fatty deposits in the body, not only about the mid-section, which is the most likely place for it to form, but in vital parts of the arteries, heart and liver. Choline, by emulsifying such fatty deposits, enables the body to dispose of them. Just as a really fine motor oil will keep an engine running clean for a long time, choline will do likewise for the human body.

Inositol

The other vital substance of lecithin nourishes the heart and provides it with regular power. It will also aid the disposal of fatty substances which have been building up in the arteries.

It is only in more recent times, that scientists and other experts, have become aware of the vital role, played by lecithin in protecting health. Lecithin in tablet form may be purchased from all health food stores. So a person who is unable to obtain or does not like such food as soya beans or wheat germ, is able to get this vital ingredient in tablet form. In this respect, it is always advisable, to insist on obtaining a supply in its natural organic form, not a synthetic product, but something pure and straight from the plant, as nature intended.

The Middle-Aged

The amazing results of the dispersal of fats in the blood of the elderly, by the use of lecithin is even more pronounced. There is a tendency for fats to remain in the blood of older persons longer. Edward F. Hewitt writes in his book, *The Years between 75 & 90*, "If lecithin is given to older people before a fatty meal it has been found that the fats in the blood return to normal in a short time, in the same way as they do in younger persons". Having heard a brief introduction into the wonders of lecithin— it is hardly surprising that the soya bean made its mark in ancient history, and is considered such a vital food today.

CHAPTER FIVE

THE SOYA BEAN IN THE FAR EAST

ALTHOUGH THIS BOOKLET is in the main dealing with Europe and the U.S.A., we must pause a little, and briefly look at events in ancient history, and today, in the East. For all this is closely related to the present day gigantic soya bean industry of America. To give an illustration, there is a company, La Choy Products, Archbold, Ohio, U.S.A. who have a market of soya foods. This company may be Chinese in origin as the name would indicate. Some of the many varieties of tinned food produced by La Choy Products include bean sprouts under the heading of:-
Shrimp Chow Mein.   Chicken Chow Mein.
Beef Chow Mein.   Chop Suey Vegetables.
Fancy Mixed Chinese Vegetables.
So we go right back to the Orient once again for food produced and canned in the States, but bearing the old Chinese names.

Cooking in China

For centuries the Chinese have valued, and considered the art of cookery to play a vital role in their lives, and to be equal in every way to all the other arts in their society. Generally speaking, people in Europe know very little about the Chinese, their way of life, and in particular Eastern Food. The opening of thousands of Chinese restaurants, and the large number of troops who visited the East in the 2nd World War, has made some amends for this. An old Chinese belief, which is just as true today as it was centuries ago, is that a man becomes what he eats. The record of survival in China during the last 5,000 years and the fact that they have the largest population in the world would indicate that they eat the right food.

In a vast continent, in fact larger than the whole of Europe, as may be imagined there are literally thousands of ways of

cooking and preparing food. However, broadly speaking all
Chinese dishes are classed under five headings—or Schools of
Cookery. Canton is one such School and practically without
exception all Chinese Restaurants in Europe and elsewhere
are Cantonese. The reason is that Cantonese cookery can easily
be adopted for international use, and the bulk of the Chinese
living abroad are Cantonese. They have a great variety of sea
foods, and are experts at preserving them by salting and then
drying them in the sun. They have introduced a greater number
of dishes than any other school. As this book concerns Soya
beans we will look at the Chinese vegetable cooking. Basically
all schools have the same ingredients. They all use soya, peanut
or other vegetable oil, not animal fat. There is no doubt that
the appetizing health giving soya sauce plays a vital role in food
preparation. They have many tasty and skilfully prepared vegetable dishes, including the soya bean and its products. Dishes
are made to resemble the shape of fowl, fish, ham, etc., and then
covered with soya bean curd skin. (This consists of bean curd,
as thin as paper, seasoned with soya sauce and dried.) Chinese
vegetable soup will resemble chicken broth in every way. Really
it is soup made from soya bean sprouts. What a fine thing for
vegetarians, who have meat eaters as friends, to be able to serve
up a meal for them which looks like meat, tastes like it, but is
really vegetable so prepared and tasty, that even an expert
would find it difficult to distinguish any difference.

Soya Sauce

Known to the Chinese as Soy sauce, it is one of the outstanding culinary inventions of the Chinese and is extensively
used, as a condiment, or to replace salt at the table. So the Chinese
automatically get a certain amount of this ancient food every
time they sit down to a meal. This sauce enhances the taste of all
food, but to obtain the right fragrance it should be used sparingly,
otherwise the taste of the sauce will dominate the entire meal.
Customers of Chinese restaurants will be familiar with the bottles
of this sauce placed on each table.

Bean Curd or Tufu

A food used extensively throughout China—a National dish to be found in palaces, or among the peasants. It is obtained by bringing soya bean milk to the boil, when a film forms on the surface, this is known as yuba. Yuba is a very valuable health giving food extremely rich in protein. In preparation for eating, it can be either dried, or used in soups, gravies or with other dishes. Bean curd is without doubt a masterpiece of the Chinese in food preparation. Although one of the cheapest foods obtainable anywhere in the world—it can easily be produced as a delicacy at a banquet fit for a king. Bean curd is so universally accepted that every Chinese School of cookery has its own special way of preparing it.

Bean Sprouts

Sometimes referred to as shoots! Two kinds are used in China, those of the soya and mung beans. The latter known as green peas. These two vegetables are unique, because of the fact that they can be grown anywhere in a few days, throughout the year, given the necessary time and attention of which the Chinese have both at their disposal. These shoots are delicious and very rich in vitamin C—served with meals in Chinese restaurants as Chow Mein or Chop Suey. These sprouts are eaten by the Chinese as a green vegetable and served as a salad.

Soya Bean Milk

Records show that this was used long before the birth of Christ. This may be consumed fresh, in the same manner as we use cow's milk. A vegetable cheese can be prepared from this milk, in much the same way as the farmer obtains cheese from cows' milk. This is very similar to bean curd or tufu previously described.

Candied Beans

This is when the beans are boiled as a sweet in syrup. They

can also be soaked in water and then roasted—very similar to roasted nuts.

There are many other and varied ways in which the Chinese make use of the bean but this brief account shows the main ones.

Japan

Many of the uses of the soya bean in China apply equally to Japan. However, the Japanese crop is only a fraction of that grown in China.

Natto and Miso

These two foods are famous in Japan and used everywhere. Both are obtained from the soya bean by a process of fermenting and steaming. A variety called Red Miso is produced by using soya beans, together with a large quantity of salt, and after preparation it will keep for many years.

Japanese Soya Sauce

Is obtained by fermenting beans with parched wheat.

CHAPTER SIX

AMERICAN INFLUENCE ON THE SOYA BEAN

ALTHOUGH THE HISTORY of the Soya bean goes back over 5,000 years in the Far East, it is only a fairly recent crop in America. The last 25 years have seen a phenomenal increase in acreage of these beans in the United States. It took the Americans a long time to realize the crop had tremendous possibilities, not only as food for the home market, but as an export crop. Later, scientists proved that it was very useful, and equally versatile for industrial uses. Gradually, they began to appreciate why the crop had been classed as a 'sacred grain' by the ancient Chinese. Cultivation progressed rapidly and China slowly lost her lead as being the world's largest producer to the Americans. By 1963 over 60% of the total production of soya beans in the world were grown in the United States.

By 1960, China with the largest population in the world, was growing only about half the soya bean acreage of the United States. A further disadvantage from the export point of view, for the Chinese with 700,000,000 mouths to feed, was that the bulk of their crop was no doubt earmarked for the home market. So the Americans had two distinct advantages over China:
1. On being able to cultivate twice their acreage of soya beans.
2. Being the leading industrial power in the world, and with a much smaller population, they were able to utilize much of their crop for export purposes. What a wonderful job they have done in putting the soya bean on the map for Western civilization.

Culture and Development of the Soya Bean in the U.S.A.

These beans have been grown in the United States for well over 100 years. Tonnage of two crops for 1920 and 1961 will illustrate how cultivation progressed bringing America into the world lead:

In 1920 there were only 190,000 acres under cultivation.
By 1961 this had increased to the enormous crop of 26,000,000
acres. Soya oil ranked as the largest among all the oil seed
crops of the Western Hemisphere. The bean became fourth
in the cash crops of America. This must be, without exception,
a record expansion of any crop throughout the world. This is
really remarkable, when we consider that prior to 1940, in the
States, the bean had only been mainly grown for hay, or used as
a green manure. The crop showed that it had many unusual
features. Not only was it very rich in both protein, oil and other
essential vitamins, minerals, as outlined in a previous chapter,
but it would thrive on indifferent soils. Another great advantage
of the crop was that it had the attraction of being almost
immune from disease.

The Bean and Other Crops

Between 1949 and 1960 the total United States acreage of
the soya bean increased by the large amount of 124%. The following figures illustrate what an important crop this is compared
to other cultivations. During the same period 1949—1960, Wheat
growing fell by 31%, Oats by 28%, Corn acreage by 4%.
Although we are interested, in this book, in the food value of the
soya bean it is of interest to note that during the same period
the cotton acreage fell by the very large amount of 44%.

Diseases of crops

'Effects of diseases on soya beans have been considered in the
past to have been of little importance' (Piper & Morse, 1923).
Until recently there was little information on disease damage to
crops. It is of interest, that most of the diseases of the soya bean
do not kill it, but simply reduce its ability to crop. Some 12
diseases are known but, as stated, they seldom cause termination
of growth.

Insects and Spider Mites

The crops may be attacked by a variety of pests but usually

they do survive. However, occasionally certain insects such as caterpillars have been known to devastate an entire crop. Certain approved chemical fertilisers are used to control such beetles and caterpillars that may cause damage.

American Agriculture

To stimulate interest, and encourage farmers to grow crops of soya beans in the United States, the government has for some years guaranteed them a certain price per bushel grown. This has proved more than successful, to such an extent that more than one billion bushels were grown in 1968. An even bigger crop is forecast for 1969, but this might well be limited by the fact that an oversupply of beans has been cultivated. As a result, the guaranteed price per bushel has now been reduced to prevent a surplus being produced.

The enormity of this crop in America can be imagined when we consider that in 1961, 26,000,000 acres of soya beans were under cultivation. The 1968 and 1969 crops of over a billion bushels each year means that there were 60 billion lbs. harvested in 1969.

Other Sources

In 1936, the United States and Canadian Departments of Agriculture co-operated in the growing of and experiments with soya crops. However, by 1960, Canada produced the lowest annual crop among the 8 nations growing the bulk of the world's supply. At this time the following countries cultivated almost the entire world supply between them—shown in order of the amount produced: United States, China, Korea, U.S.S.R., Brazil, Canada and a very small amount produced in various other places.

Disease Resistant Varieties

Much progress has been made in recent years in the U.S.A.,

where they have produced new strains of soya bean crops. These are disease-resistant and heavy croppers.

Influence on American Food

The intensive growth of this crop in the United States brought with it many changes in eating habits. The bean and substitutes of it, were incorporated into a wide range of foods, especially when people began to realize how healthy soya foods really are.

The Soya Bean and the American Housewife

In dealing with a food crop that for centuries has been a staple diet of the East, there is bound to be some overlapping in menus prepared in the Orient and in the Western World. A case in point is the American firm La Choy Products, who market all manner of Oriental tinned Chinese food from a crop grown, prepared and canned in America. This is mentioned in Chapter Five. In the main, it is a question of adopting an Oriental vegetable to suit American or British eating habits and tastes. Like the Chinese, the American wife is able to prepare all manner of appetizing meals from the bean and soya products. These range from snacks to additions to every course in a meal, from soup to sweet: nutritious food which will satisfy the hunger of all, from growing children to the elderly. Very sustaining in the diet of athletes performing strenuous physical activity, or those carrying out heavy work. The Americans are able to produce a series of courses, fit for a banquet, and not one will taste the same.

Meat and Vegetables in One Crop

The soya bean. Christened 'The Meat of the Soil' 5,000 years ago by those wise old Chinese farmers. Proved in the 20th Century by American scientists. Many organizations in America have been experimenting in the manufacture of ersatz foods. The astounding results are: Chicken, turkey, pork, duck, ham and so on—you name the meat dish, they produce it—as simple as that. These reproductions of our Sunday roast or Christmas turkey are mainly constituted of soya bean proteins.

There is an estimated fifty million or more Americans, who, for one reason or another prefer a meatless diet. The future looks very rosy for them. Very soon they are certainly going to get tasty, meaty, soya foods as a replacement—and the remarkable thing is that from the taste they will not realize they have changed to vegetarian food.

Milk and Milk Substitutes

Not content with meat and vegetables growing on the bean-stalk, Americans are now able to obtain a wide range of soya milk products. These include ordinary milk, cream, cake fillings and ice cream. However, as yet soya milk does not taste exactly like fresh cow's milk. It rather resembles reconstituted milk. No doubt in due course, with an abundant bean crop at their disposal, the Americans will produce a soya milk which is identical to our dairy produce.

Other Products

The American housewife, in addition to these rather astounding foods from the bean has:

Soya flour

This can be used for the making of tasty health giving bread or cakes. Also to add protein and goodness to ordinary flours, soups and gravies.

Bean Sprouts

Similar in every way to those produced by the Chinese and Japanese, and outlined in a previous chapter. The Americans have an extensive canning and freezing industry to deal with much of their bean produce.

Soya flour, grits, flakes used by the Americans for cake, bread and famous doughnut making. They also utilize bean products

for the purpose of making noodles, breakfast cereals, macaroni
and various health foods. The bean is roasted like a nut, and used
as a sweet after being cooked in syrup. A high percentage of the
oil of the American soya crop is used for margarine, salad and
cooking oils.

American Soya Bean Association

Founded in 1920 to protect the interest of soya bean producers
and handlers. A most efficient organization bringing together all
connected with the industry in any way. The main objects of
the Association: to safeguard the future of the crop and promote
markets, a monthly magazine *The Soya Bean Digest* and an
Annual Blue Book of Soya Beans are published.

The headquarters of The American Soya Bean Association is
at Hudson, Iowa, U.S.A. A hive of activity exists at the executive
office where many thousands of queries are answered annually
about every aspect of the soya bean crop.

Educational Films

A number of coloured films are used for training and educational purposes showing the cultivation of the soya bean at
every stage. The association has these films for the use of growers,
distributors, soya protein food industries, and all others interested
in any way in the soya bean, including school children.

Japan

In 1956 The American Soya Bean Association made an
agreement with Japan to establish a Soya Bean Market Development Project. This came into being with the formation of the
Japanese American Soya Bean Institute in Tokio. The institute
has done a great deal to promote all aspects of the vast Soya
Bean Industry. Japanese trade groups have visited the U.S.A.
to see for themselves how the industry works. Trade Fairs in
Japan have been sponsored where U.S. soya beans and all the

bean products have been exhibited. There has been a tremendous increase in the consumption of soya oil and other products in Japan since the institute was formed. Also the protein content of the crops have been used more by the Japanese in group feedings in factories and offices.

Soya Proteins in Foods

Forgive repetition if I again quote that, 'only a minor part of the protein of the American soya crop is used in human foods' at present. The consolation is that this percentage is steadily increasing. This has been brought about by the remarkable way soya protein can simulate all meat dishes. This is indeed exciting news for the average American, most of whom are ignorant of the soya protein meats industries as the man in the street in England. For many years now in the U.S.A. soya proteins have been extensively used as a low per cent. additive to a wide range of foods including: biscuits, cereals, muffins, bread rolls, cakes, frankfurters, sausages, meat patties, meat loaves, stews, gravy, poultry, soups, puddings, candy, confection, baby foods and other items.

These proteins are used in a greater percentage in health foods.

Advantages of Soya Proteins

These protein additives have many great attractions. They are very nutritious, containing all the amino acids vital for good health. Being moisture retainers, emulsifiers and binders, they have many other uses. Shelf life of such products is greatly increased. Appearance of food is improved and processing costs reduced. Finally soya flour, grits and meal contain a rich supply of B complex vitamins, choline, calcium, iron, phosphorus, potassium and traces of other essential minerals.

Types of Soya Proteins

Some soya proteins are as old as the bean crop, being a food

fish, fowl, and even nuts. An increasing demand is being made for Worthington 20th Century foods, but as yet, consumers are mainly vegetarians or others who for health or religious reasons seek a meatless diet. However, there are signs that a change is gradually being made in American eating habits. This as because scientists and individuals have shown that human beings can not only live but remain remarkably healthy on diets without animal protein. More people will buy these foods as they become aware of them and realize how good they are.

Many Attractions

These foods are free from cholestrol, fat and diseases sometimes found in animal protein foods. They contain no waste and remain constant in taste and are tender. Being mostly precooked they make for a revolution in the kitchen. They can be marketed frozen, dehydrated, freeze dried or canned. In part they may be produced as ingredients for other foods such as pizza, or as meat sauce for spaghetti or other dishes. As mentioned, some of them are as cheap to buy as meats and as they become more popular prices are bound to go down.

Future Prospects

For the past seven years Worthington Foods sales have increased by some 20% annually and it is considered there will be an ever greater expansion. It is interesting to examine the reasons why this company are confident their sales will continue to increase. They are:

1. These foods are nutritious and taste good.

2. Vegetable proteins are more economical to produce than traditional meats.

3. No wastage and a minimum amount of work is required to prepare these foods in the kitchen.

4. There is a world wide shortage of protein which, as the years go by, will get worse with population increase and people living longer.

So if we can accept these reasons it would seem the soya protein foods companies have every right to be confident in their produce.

CHAPTER SEVEN

INDUSTRIAL USES IN AMERICA

SOME DETAILS OF the food value and uses of the soya bean and its numerous products as an aid to health have been given. From this it will be seen that as a complete protein and protective food, the soya bean rates as second to none. We now take a look at the role, an ever increasing one, that it is playing in the world of commerce.

Man, as always has not been slow to exploit this crop to serve his purposes in Industry. In Western society today many persons are far more inclined to take a greater interest in cars, home decorations, and bigger and better gadgets, than in good health and wholesome food. The versatile little bean has aided them. However, it is only fair to point out that the Americans, who have made a tremendous gift to the world in producing today some billion or more bushels annually, utilized up till 1963, only 11% of the entire crop for industrial purposes.

Paint Manufacture

For many years the soya bean oil has been used for making excellent paints. These include white paints, used for domestic purposes in kitchens, and all manner of coloured paints for use in industry. This includes the spray painting of cars, machinery and many other allied uses. Soya bean oil has also been utilized for making resin paints and varnishes.

Printing Industry

Here the oil finds its way into the processes for the manufacture of printing inks, adhesives, paper and glues. With an ever increasing demand for more papers, periodicals and books this use is likely to be increased.

Brewing

In the U.S.A. the bean is used in the brewing industry to provide nutrients to aid and assist the growing of yeast. By this means the flavours of many beers have been enhanced. This particular use of the bean may well be considered to be equally as much towards food production as industrial.

Plywood Industry

The remarkable adhesive power of soya glue has made a tremendous impact on the manufacture of plywood in the U.S.A. In fact, up to 1963, the total tonnage of bean glue used exceeded that of any other adhesive in the industry.

Pharmaceuticals

Once again, here we find that the oil of the bean plays an increasing role in the production of soaps, cosmetics, and other aids to beauty in the chemist's shop.

Other Uses

These include productions of various oils. Making of sealing and caulking compounds. In the manufacture of linoleum and oilcloth, plastics and textile fibres. Another use is in the making of fire appliances, where, it has been found the oil acts as an emulsifying agent.

No doubt, at the time of writing this booklet, more and more ways of using the bean to serve man in the 20th Century are being discovered. If the commercial uses of this vegetable ever equal the food value, then it will become an industrial wonder of the world.

There is little doubt that the soya bean is destined to play an ever increasing role, both as a food and for industrial purposes. Research and experiments continue, in which more use than ever will be found for this versatile crop. Little did those ancient Chinese realize what part their humble little bean would play, in the commercial world of the Atomic Age.

Summary

A comparison of the disposal of the entire American Soya Bean Crop for the year 1963, shows that 89% of the soya oil was used directly in human food production as hereunder:

> 31% used in the making of margarine, cooking oils, salad oils.
>
> 34% For the purpose of making shortening.
>
> 24% Other food uses.
>
> ─────
>
> 89%
>
> ─────

Unfortunately, a totally different story must be told of the use made of the protein content of this particular crop. In fact, 95% of the protein was directly used in animal foods.

Thus, in spite of intense industrial development, only 11% of the oil crop was used for purposes other than food supplies. It does seem unfortunate that so much of the protein could not be adapted for human consumption.

CHAPTER EIGHT

THE SOYA BEAN AND WORLD FOOD SHORTAGE

THE AMERICAN NATION has rendered a valuable service to the entire world in cultivating a gigantic soya bean crop. From an experimental stage during 1936, rapid progress was made as shown:

14 million acres were harvested in 1950.

Eleven years later, 27,000,000 acres were grown.

In just over ten years they had practically doubled this crop. If the same miraculous intensive cropping of this bean can be carried out in other countries, with a suitable climate, there is more than a gleam of hope for the world's hungry people.

It is perhaps only to be expected that the most wealthy nation in the world, with millions of acres of land, much of it in an ideal climate for bean cultivation, should do so well with these crops. What is perhaps not realized is that in the 21st Century, millions of lives, hanging in the balance through food shortage, could well be saved by the export of American soya products.

Breakdown of American Soya Crop—1963

A study of the American crop for 1963 shows that nearly 90% of the soya oil produced was used in foods for human consumption. It was converted into margarine, shortening, cooking and salad oils. At the same time, 95% of the protein of this particular crop was put to animal feeding. At a time when the population of many parts of the world were underfed, it seems unfortunate that such a vast tonnage of body building protein foods should be used as animal feeds. Unfortunately the position has not changed for in 1967 only about 2% of the protein of the United States soya bean crop was being directly utilized in human diets. This is in spite of the greatly renewed interest in using soya products to alleviate the world's nutrition problems.

Without doubt there are many other problems involved. Can America export a far greater percentage of her soya bean crops for human consumption?. Obviously up to 1963, there was not such a great demand for export of the protein element of the crops, as is the case with the oil content. No doubt large tracts of farmland previously used for the growing of hay and other cattle feeds have been converted to soya bean culture. Growing the bean for the farmer serves the twofold purpose of:

1. Producing a major oil crop for world wide export, and use on the home market. Soya oil is the largest oil supply of the western hemisphere.

2. Having an even larger protein residue which can be utilized for cattle feeding. (Soya crops average 40% protein—20% oil).

Distribution of the American crops since 1963 have no doubt produced a change in export figures. As more countries are becoming aware of the value of soya bean protein, a greater demand for it is bound to follow. Also, in the United States, a number of protein food companies have been formed to produce and export such items as soya milk, flour, and protein meats. This is dealt with in greater detail in Chapter IX - under *Recipes*.

The American Soya Bean Culture and the industries that are rapidly developing with it can look forward to a great future. There will be an increasing world demand for all these exciting health giving new foods.

Expanding World Population

In practically every country the birth rate is going up. In Western countries people mature earlier, marry younger, and live longer. There are more persons over the age of 65 than ever before, and this figure is bound to increase. While disputes and differing opinions are held over the use of contraceptives and legal abortions, the population of the world continues to increase at a most alarming rate. At the same time, in many countries towns are growing and agricultural land is being swallowed up. It is considered that an ever increasing problem will be caused in trying to feed the masses of people.

Change in Eating Habits Necessary

Vegetarians are the people who are going to reap immediate benefit from the growing soya protein industries. Likewise, such races as Hindus, who have always abstained from eating meat on religous grounds. However, it is not a simple matter to get people who have been meat eaters all their lives to make sudden changes to vegetable—meat protein—diets. Many would not be persuaded by the fact that the new kind of meal resembles the old in taste. If people are faced with a food shortage, or eating synthetic meat, then there is little doubt they would soon get used to the new fare.

So one of the greatest obstacles to overcome will be to sell the idea of meat, milk and vegetables all from one crop of beans. For centuries, man, in many parts of the world, has accepted meat and poultry as being the mainstay of his diet. It may equally be thousands of years before a real noticeable change takes place. At present the protein meats are still quite expensive. If the industries flourish, prices are bound to come down to a figure well below meat prices, which are rising all the time. When this does in fact come about it could be a deciding factor in many countries. Only time will tell. So far there seems to be widespread ignorance, of even the existence of the soya bean as a crop, and most people would be more than surprised, if told that a protein meat could be obtained from beans.

Future Prospects

There is a great demand for soya oil. It has the largest distribution of any food oil in the Western hemisphere. Likewise, the potential exists for selling the protein, for human consumption from this vast crop. Once prejudice has been overcome, and protein meat prices are able to compare favourably with those of fresh meat, the future should be bright for these newly formed protein meat companies.

Booming Soya Milk Industry

Not long ago the suggestion of a milk supply from beans would have been ridiculed as a fairy story. Today, the story has come true for the Vita Soy company of America, who market millions of bottles of this milk throughout countries in the Far East. Soy milk has two great attractions, it is cheaper than cow's milk in the countries where it is sold, and secondly it is very nutritious. Richer than cow's milk in iron, calcium and phosphorous. It also has the added attraction of being much easier to assimilate than ordinary creamy cow's milk in a warmer climate.

The success of the 'Soy Milk' Company indicates that the markets are available, and can be exploited to the full—for the benefit of everyone concerned. In time Soya Meats will follow the same trend.

Many Varieties

The Worthington Foods, Worthington, Ohio, U.S.A., have also gone into the Soya Milk business in a big way. They produce a milk powder called Soyamel, of which there are many kinds including, Soyamel Banana Beverage, Malt Soyamel, Fortified Soyamel and Regular Soyamel. The essential food elements in Soyamel include protein, fat, carbohydrate, calcium and phosphorous. One quart of reconstituted Soyamel (which dissolves instantly in water) and takes about 4 ozs. of powder to make provides the following nutritive values:

Vitamin A	5,000 U.S.P. units.	Naicinamide	15mg
,, D	400 ,, ,,	Calcium	750 ,,
,, E	10 ,, ,,	Phosphorus	750 ,,
,, C	100 mg.	Copper	1.5 ,,
,, B1	2 ,,	Manganese	2 ,,
,, B2	3 ,,	Zinc.	5 ,,
,, B6	2 ,,	Iron.	15 ,,
,, B12	2 micrograms.	Iodine.	100 micro-grams.

It will be noted that all of these essential bodily needs have been mentioned in Chapter 3, *Content of Soya Beans*. So it is interesting to find out that soya milk is equally as nutritive as the fabulous bean. One quart of Worthington Fortified Soyamel is one of the most health-giving drinks in the world. This amount consumed daily provides all the vitamin and minerals essential to health for a normal adult, in addition to the protein content and essential amino acids. The outstanding quality of Soyamel is its flavour, being free of the characteristic bean taste sometimes present in soya beverages. This milk looks good, tastes good and is good. Being very versatile it is equally suitable for cereals, vegetable soups, cooking, baking and all the other purposes for which cow's milk may be used.

Plamil or Plantmilk

Supplies of soya milk powder are not obtainable in England. However, one company is actually producing what is in effect soya protein milk, in a concentrated liquid form. It is supplied to health food stores in ¾ or 1½ pint tins. Water is simply added and it is ready for serving. Containing soya protein, this product has all the advantages of cow's milk and is a good source of lecithin, vitamins A, B, B12, D and calcium. It is a pure food drink being free from animal fats, artificial colouring and preservatives. So although it may be some time before powder soya milk is produced on a large scale in England, the liquid soya protein milk is becoming increasingly popular, and proves ideal for persons unable to take cow's milk, or others who have been advised to reduce their intake of animal fats.

Much may be said, for and against, the development of the Soya Bean Protein Industry. Man will insist in having, whenever possible, a freedom of choice. Thousands may never become resigned to the idea of living on a vegetarian diet, even if the protein is equal in all respects to animal flesh. On the other hand the majority of confirmed vegetarians would never change to a meat diet. However, with all the facts and figures, and arguments for and against, the introduction of 'meat of the soil', only man can decide for himself.

There is one vital and what may prove a deciding factor in favour of our Soya Meat Bean Produce. The average amount of protein which can be harvested from an acre of soya beans is nearly six times as much as that gained from an acre devoted to farming livestock.

This indeed is a comforting thought in times of inflation and ever increasing prices. Meat costs, together with all food prices, increase slowly but surely, and there is no hope that they are ever likely to come down again.

Soya bean farming in America is a new crop—much has been learned about it, and more and better yields of beans are being produced. There is no question but that further improvements will follow. In many areas The Giant Yellow Mamoth bean has been replaced with a bean known as the Lee Soya Bean—this is said to be almost immune from disease and other setbacks. So progress continues in this new and fascinating industry.

Growth of World Soya Bean Crops in 1960

The United States and China lead the world in cultivating soya beans. The American crop was about double that of China.

Then followed countries as shown in order of priority of acreage grown:

> Indonesia
>
> Japan
>
> Korea
>
> U.S.S.R.
>
> Brazil
>
> Canada

The remainder of the growers in the world produced between them an acreage as large as that of the lowest country on the list.

Although China grew the largest crop after America, they also have the biggest population in the world, about 700 million. With such a massive population to feed it is doubtful if China is able to export very much of ther crop. The remaining six countries cultivated only a very small fraction of the total world crop in 1960. Roughly, America produced twice as much as China, and all the other countries in the world who grew the crop, barely cultivated together a quarter of the crops grown in China. So it would seem that these countries, too, would not have a very large export market. The American soya bean crop has continued to grow and the forecast is that it is likely to do so for many years to come. In March 1969, some 41,000,000 acres of American farmland were under soya bean cultivation.

Summary

There is little doubt that a far greater world acreage of the soya bean could be cultivated. Such countries as Australia, New Zealand, and many parts of Africa have ideal climates.

Canada took a great interest in the crop, together with America, and carried out experiments and planting of species as far back as 1936. However, the Canadian Government do not seem to have made great progress. Possibly some parts of Canada are too far North for the crop to be grown.

The first priority would seem to be to introduce the soya bean to the world's communities. It is a very new crop to the West, and America has certainly performed a formidable task, in the past 20 years, in developing the cultivation. In the States they have also made rapid strides in expanding the Soya bean industries.

As mentioned previously in this booklet there is a tendency to overproduce the bean in America. Farmers receive a subsidy from the American government for growing the bean. In order to counteract the over-production trend, farmers' subsidies in the States will be reduced in 1969. No country in the world, the USA included will be prepared, or can afford, to cultivate soya beans, or any other crop unless they are sure of a market.

Soya Oil

In the U.S.A. and many other countries soya oil has been accepted for a long time. The bulk of the oil used in margarine shortening, salad and cooking oils is soya oil. It is of excellent food value being rich in lecithin. The big problem, of course, is for producers to try and obtain like results with the protein content of the bean. No problem will exist here for vegetarians or sections of the world's community who for religious or other reasons abstain from, or curtail their intake of animal flesh. All that is necessary is to bring knowledge of the bean and the soya protein foods to them and show how vital they are to fill gaps in meatless diets.

Soya Protein

When the demand for soya protein becomes as great as that for the oil the full benefits of this wonder crop will be available to man everywhere. For various reasons nowadays more people than ever are wondering if a daily intake of animal flesh is so vital to health. Meat is becoming increasingly more expensive to buy. Carcases of animals are found to contain greater amounts of DDT, and antibiotics are increasingly used to treat livestock. All these facts cause some people to purchase less meats.

Soya Milk: Milk Without Cows

For over 5,000 years this health-giving soya milk was part of the food of the ancient Chinese and other Eastern races. It is made up of an emulsion of soya beans in water. In 1959 The World Health Organization constructed a 1,000,000 dollar Soya Milk Plant in Indonesia. Many million bottles of this milk are produced yearly by K. S. Lo. of Hong Kong. Very nutritious and cheaper than cow's milk, it has helped many of the unfortunate people in the East to survive. In the Western world Soya Milk has also been used to combat malnutrition or for persons allergic to cow's milk.

Grave Problem of the Future

The entire world animal agriculture is faced with a tremendous task. If the estimated population of the world in 1975 is to be allowed the same percentage of meat as is available today, production must be increased by some 30%. A formidable task indeed. This is where the little soya bean and all the amazing Soya Protein Foods will prove more than a godsend —the difference for millions between enough to eat and starvation. This becomes apparent when I again quote: '*The average amount of protein which can be harvested from an acre of soya beans, is nearly six times as much as that gained from an acre devoted to farming livestock*'.

The Hunzas

Is a daily intake of meat necessary for health? Obviously vegetarians and other non-meat eaters will say that it is not. One remarkable race of people exist on a practically meatless diet and live long useful lives to a great age, free from many diseases. These are the Hunzas who occupy a fertile valley, one hundred miles in length in West Pakistan. Fifty years ago a Dr. Robert McCarrison, Director of Nutritional Research in India spent seven years with the Hunzas. In the book which he subsequently wrote Dr. McCarrison has this to say about them:

'A race unsurpassed in perfection of physique and in freedom from disease. Amongst these people the span of life is extraordinarily long'.

An American, Rene Taylor, is one of the most recent writers to visit the Hunzas. Her book, *Long Suppressed Hunza Health Secrets* became a best seller. She met Hunza men and women enjoying robust physical and mental health at one hundred years of age and over. It is accepted that Hunza men of ninety and one hundred years of age can father children. She saw people of eighty and ninety in a perfect state of health who had every appearance of being in their forties. Cancer, heart disease, and other serious complaints, accepted as unavoidable in Western countries were practically unknown among the Hunzas.

Diet of the Hunzas

Naturally there are many factors which contribute towards the long lives and robust health of these people. Their relaxed way of life is free from the stresses and pressures of civilization, for they have no prisons, banks, money taxes, stores. There is no crime, no divorce and education is free to all. They practice their own kind of Yoga, can walk for miles and lead strenuous outdoor lives. With all these aids to good health, even in the beautiful Hunza valley, in the long run what the people eat greatly determines what they become. Let us take a look at the diet of the Hunzas in the 1960's as Rene Taylor saw and wrote about in her book, one of the chapters of which is headed, 'The Amazing Hunza Diet'. Food which has provided centuries of good health for these happy, carefree people consists of grains, including wheat, barley, buckwheat and small grains—only wholemeal, white bread is never eaten. They eat leafy vegetables potatoes and other root crops. Rene Taylor found that the soya bean played a vital and important part in the Hunza diet, providing protein for the making of cheese, milk, bakery foods, etc. She said they only ate meat on rare occasions such as holidays, weddings and so on. Among other vegetables, sprouted soya beans were very frequently consumed.

There must be a number of reasons for the amazing vitality and long lives of the Hunzas. Being almost vegetarians they might well be expected to experience some shortage of protein. Of course there is not the slightest indication that they suffer from any such deficiencies. The little soya bean consumed extensively by this healthy race of people without doubt plays a vital part in keeping them not only strong, healthy and free from disease, but puts them on their own as one of the fittest races in the world.

It seems obvious if more people become familiar with the soya bean, and accept it as part of their diet, greater cultivation will surely follow. Without doubt it is a crop of the future for the Atomic Age.

The world and its teeming millions are bound to feel the effects, in the years ahead if the cultivation of this vital food is not allowed to keep pace with the ever increasing population. America has done her share, and it is hoped that other countries, who are able to cultivate this remarkable vegetable will make every effort to do so.

CHAPTER NINE

SOYA BEAN RECIPES

Some of these recipes for the bean have been handed down through ancient Chinese history, others are as up-to-date as the 20th Century Soya Bean Processing Industries. However, it must be remembered we are dealing with a vegetable which originated in the Orient and is new to Western civilization. For centuries this crop was grown exclusively in China, Japan and other Eastern lands. Therefore, it must be accepted that many basic ways of preparing and serving this ancient food are predominently Eastern in origin. *Only biochemic or sea salt is recommended for use in preparing any recipes from soya beans or bean products.*

Gradual switch to human consumption

It is only 100 years since the first species of the bean were grown in America. For the next 50 years or more, it was just another minor crop produced for good farm manure, hay, and protein animal food. Twenty-five years ago the cultivation was expanded, and during the past ten to fifteen years a most remarkable increase has taken place of soya bean cultivation. With the growth of this crop, various ways of using the bean and its products for human consumption have emerged.

We have the Americans to thank for providing the Western world with the soya bean. Progress was very slow at first but in 1963 the Soya oilseed crop produced more oil than any other cultivation in the Western Hemisphere. These oils are used for the purpose of making margarine, shortening, cooking and salad oils. The Americans put everything into the development of the crop. In a few short years their success was such that the world lead passed from China to the United States, and the Soya bean rated fourth as a cash crop in America.

Slow Progress from the beanfield to your table

Unfortunately, with the exception of the Soya oil Industry, progress in utilising the bean and its protein content for human food is very slow. Considering the size of the crop, only a fraction of the valuable protein ever reaches the table. Almost the entire oil production, about 90% of the total crop, is used for human food as outlined in the previous chapter. Up to 1963, the percentage figures were almost reversed for use of the protein content of the crop, when some 5% was used for human foods and drinks, and the massive amount of 95% went direct to animal feeding.

Growing Soya Bean Industries

A percentage of the crop, exact amount not known (by the writer), but in 1963 it was estimated that 5% goes into the manufacture of various foods as outlined below:—

Soya beans direct from the fields used as fresh or dried vegetables.

Soya Bean Milk. This a rapidly growing business, and America is able to supply canned milk to many underdeveloped countries at a much cheaper rate than cows' milk.

Soya Flour. Also a growing industry. It is produced by the milling of beans in Europe and many overseas countries. This flour is more nourishing than any other kind. Unfortunately, it costs more than ordinary flour at present.

Breakfast cereals. Mostly produced from the bean in the USA for their own market.

Confectionery and Ice Cream. This is yet another field in which the bean is being used, mainly in America.

New Food Industry. Newly formed companies in America producing protein 'meats' from the bean are demanding their cut of the Soya cake. So a certain slice of the crop, an ever increasing one, will soon provide the entire vegetarian population of the States and elsewhere with Meaty Vegetable Foods, straight from the bean crop. The American housewife of the future will be able to obtain her weekly joint of beef,

pork, fowl and even the Thanksgiving Turkey from shops
supplied by Protein Producers—the soya bean in another of its
varied forms.

Sausages, Meat Pies and sliced Meats

For some time now soya bean proteins have been used by
firms in America, together with natural meats, to make up frankfurters, hamburgers, meat pies and so on. The percentage of
soya protein used may range from 10 to 25%.

Gigantic Food Industry of the Future

The use of the bean for most of the above industries is something very new. People do not always take kindly to such changes,
and no doubt it will be quite a long time before the average
Englishman can be convinced that the roast beef, which he eats
on Sunday with Yorkshire pudding, can come from a row of
beans. However, the use of soya crops for these purposes is an ever
expanding one. It may eventually take its place with the important
Soya Oil Seed Industry of America. Providing, of course, ways
and means can be found of getting sufficient animal feeds from
other sources. It can be said, that if millions of meat eaters do
gradually change over to soya protein food and milk—then we
will not have to keep anything like the present herds of cattle,
for either meat or milk consumption.

Recipes in General for everyday use

Soya Sauce. Eastern meals are flavoured with what is called in
Japan Shoyu sauce or Soy sauce in China. It is served at practically every meal, in the same way as we use condiments. Soup,
fish, poultry, meat and vegetables are flavoured with it. Cold
dishes of fish or meat are likewise flavoured with this sauce.
Only a small amount of sauce should be used, otherwise it will
predominate the entire meal. It is made from a mixture of soya

beans, flour, salt, sugar, and caramel, and has a pleasant and distinctive flavour. Bottles of this sauce may be purchased at any delicatessen shop. It is certainly not worth the trouble or expense of trying to make this sauce (even if we are able). A bottle will last quite a long time, even with daily use, for it is very strong and concentrated, only a few drops are needed at a time. Worcester Sauce is only made from soya beans.

Tofu. Often known as Dried bean Curd, is another famous Japanese food from the bean. Cubes of this curd are used in Japanese cooking. The beans are soaked in water for 1-2 days. They are then pounded together, and put into square moulds with brine and boiled till they become hard. This again is hardly the kind of recipe one would be prepared to make in this country.

Bean Sprouts. Soya bean recipes would not be complete without a description of the making of bean sprouts. They have always been extensively used in Japan and China alike, and are becoming increasingly popular in other countries. To produce about 2lb sprouts. Soak a cup of beans in water for 2-3 days until such time as they have commenced to sprout. Then place all the sprouted beans in a large plant pot. Cover with a damp cloth. Pot should be kept in the dark in a temperature of about 70 degrees. It must be watered frequently (every few hours) never let it become dry. In about 7 days the pot will be crammed full of delicious crispy bean sprouts which are ready for eating. These vegetable sprouts can either be cooked or served as a salad after the roots have been removed.

Preparation. The correct way of cooking the sprouts is by Chow (this means in a small amount of oil, prepared quickly, the sprouts being stirred in the oil whilst cooking). They can be prepared with all kinds of meat or poultry, or with other vegetable courses. All other ingredients should be cut to the same thickness as the sprouts. The sprouts must not be cooked too long otherwise their fine taste and rich vitamin content is spoilt. They should be tasty and crisp when served.

Vegetable Chop Suey

Ingredients

1lb Soya bean sprouts	5 tablespoons water
¾lb onions	2 tablespoons cornflower
½lb carrots	½ teaspoon yeast extract
3 sticks celery	1 teaspoon Soya sauce
½lb cabbage	1 cup stock
4 tablespoons nut oil	Salt

Method. Peel and cut up onions, celery and cabbage. Put oil in pan and heat up. Add vegetables and cook for about 6 minutes. Put in stock. Bring to the boil. Simmer for about 6 minutes. Add sauce and yeast. Add mixture of cornflour and water. Cook for about 6 minutes. Add bean sprouts and cook for 3 minutes. Serve hot with boiled rice.

The above makes a really tasty, health-giving meal. It is an ideal introduction to Chinese food, for the Vegetarian.

Soya Bean Curd or Soya Cheese. A very nutritious, tasty cheese—fresh soya beans should be used. If these are unavailable, dried beans may be left in soak overnight and used.

Ingredients

½ to 1lb fresh soya beans (or dried   Little milk, yeast and salt
    as the case may be)

Cook the beans in water without salt. When they are nearly cooked take off most of the water. Continue the cooking for a while till the skins are tender. Empty contents out into basin (earthenware) and leave overnight. This will allow the contents to become jellied. Reheat and drain off the rest of the liquid. Pass all the beans through a fine sieve. Add to the puree the drained off liquid. Add little milk and some yeast to produce fermentation and salt to taste. Divide preparation into cheeses of sizes required and leave them to 'take' when they are ready for eating.

Bean Curd with Fish. This makes a very fine dish.

Ingredients

Peanut or soya oil	Soya sauce
Bean Curd and Fish	Spring onion
1 glass sherry with a little green ginger	Salt

Fry the piece of fish (whole) over low flame until it is brown on both sides. Pour over the sherry and ginger mixture. Add tablespoon of soya sauce and chopped spring onion, salt to taste. Cook for further 15 minutes and add the bean curd. Serve hot as soon as curd and fish are cooked.

Beans for breakfast or with any meal. Dried soya beans may be simply prepared. After washing the beans, leave in soak over-night. They will absorb a lot of water. Simply bring to the boil and allow to simmer for 30 minutes to 1 hour—depending on the amount being cooked. Add salt to taste. The less the cooking the more nutty the flavour. They then may be served as they are with bacon for breakfast, or used as an additional vegetable course with any meal.

To add variety the beans may, whilst cooking, be mixed with finely chopped onions, tomato, little sugar, salt and spices.

A more nutritious, appetizing feast of beans will be hard to find.

Soya Milk. This again is the kind of soya product which few people would attempt to make. This milk is a very nutritious drink, used by the Chinese before the birth of Christ. In America today this milk is being produced on an ever increasing scale.

It simply consists of the mixture of cooking liquid and puree from Soya beans before fermentation starts.

Roasted Soya Beans. Beans are soaked in salt water overnight and roasted in a similar way to preparing roasted peanuts or chestnuts.

Candied Beans. After being soaked in water the beans are boiled in syrup and served as a sweet.

Baked and boiled beans. The Americans have a way of baking or boiling beans in a manner similar to the canned baked beans.
Soya Flour. Soya flour is well known to users of Health Food Stores. It has wonderful food value, milled from the soya bean and containing 40% protein. One supplier provides a free comprehensive recipe list containing nearly 40 recipes ranging from soups, pastries, bread, biscuits, potato meals, breakfast and cereal foods and a variety of all manner of sweets.